*In the same series*
KNOW ABOUT—*Transport*
ISBN 0 905778 02 2

English translation © 1978 by Richard Sadler

Illustrations © 1976 by Altberliner Verlag Lucie Groszer

First published in this edition 1978 by Wrens Park Publishing, Barton under Needwood, Staffordshire

First published 1976 by Altberliner Verlag Lucie Groszer, Berlin, under the title *Der verwandelte Wald*

Published in Canada by Carlton House (Roger Carlton Ltd), 91 Station Street, Ajax, Ontario L1S 3H3

Filmsetting by Alden Press, Oxford
Printed in the GDR by Sachsendruck Plauen

ISBN 0 905778 03 0

KNOW ABOUT— *Wood*

*Designed and illustrated by* Rainer Sacher

WRENS PARK
PUBLISHING

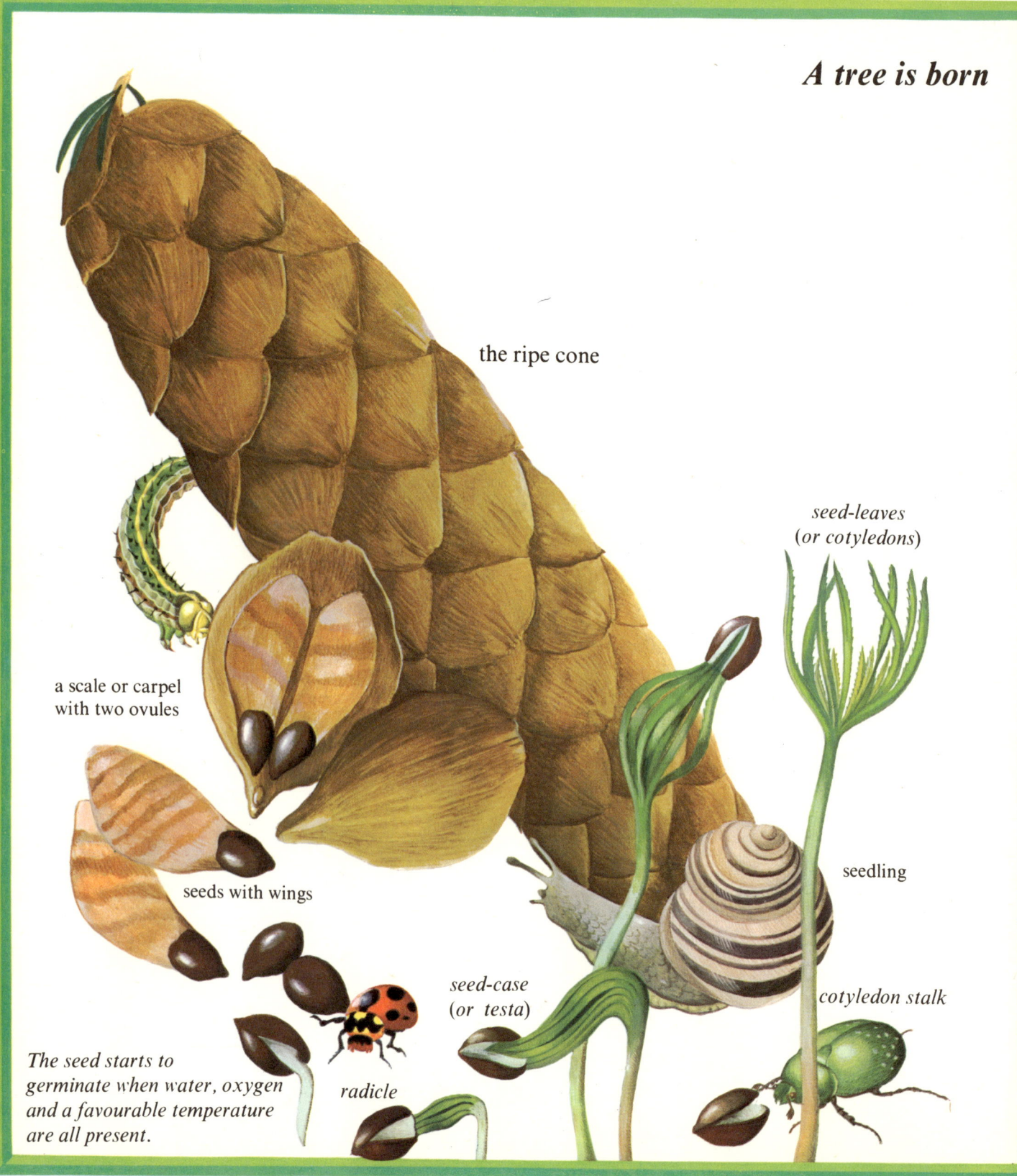

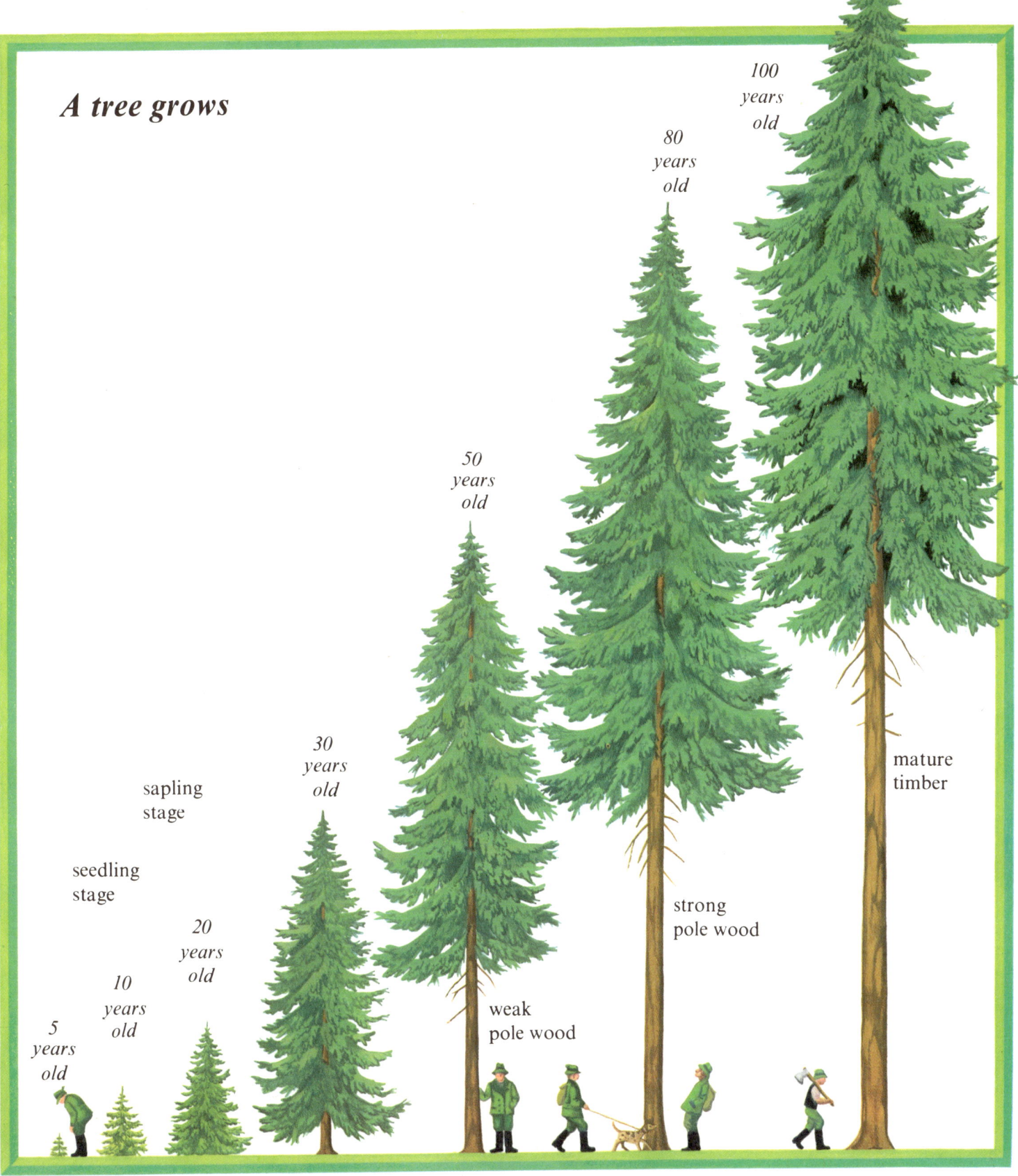